Effective Utility Management

A Primer for Water and Wastewater Utilities

Foreword

Water and wastewater utilities across the country are facing many common challenges, including rising costs, aging infrastructure, increasingly stringent regulatory requirements, population changes, and a rapidly changing workforce. Effective utility management can help utilities respond to both current and future challenges and support utilities in their common mission of being successful 21st century service providers.

Based on these challenges, EPA and six national water and wastewater associations signed an historic agreement in 2007 to jointly promote effective utility management based on the *Ten Attributes of Effectively Managed Water Sector Utilities* and five *Keys to Management Success*.

This Primer is an outgrowth of that agreement and distills the experience of a group of leaders in water and wastewater utility management into a framework intended to help utility managers identify and address their most pressing needs through a customized, incremental approach that is relevant to the day-to-day challenges utilities face. In the future, the Collaborating Organizations will continue to work collectively and individually to implement a range of short-term and long-term actions designed to promote and recognize excellence in utility management based on the principles and practices described in the Primer throughout the water sector.

We, the Utility Advisors and Collaborating Organization representatives who participated in this ground-breaking effort, believe that this Primer will be helpful to both individual utilities and the water utility sector on the whole. Based on our own experience, as well as the experience of others across the country, it is clear that effective utility management is critical to helping utilities address challenges, improve performance, and be successful in the long run. We strongly encourage all utility managers, regardless of their utility's size, budget, and unique circumstances, to read, consider, and implement the strategies and approaches outlined in this Primer.

Sincerely,

Utility Advisory Group

Cheryl Farr
East Bay Municipal Utility District

JC Goldman, Jr.
United Water

Dan Hartman
City of Golden Public Works

Mary Lappin
Kansas City Water Services Department

Ed McCormick
East Bay Municipal Utility District

Howard Neukrug
Philadelphia Water

Kanwal Oberoi
Charleston Water System

Tyler Richards
Gwinnett County Department of Water Resources

Thomas Sigmund
Green Bay Metropolitan Sewerage District

Mary Snyder
Sacramento Regional County Sanitation District

Joseph Superneau
Springfield Water and Sewer Commission

Todd Swingle
St. Cloud, Florida Environmental Utilities

Diane Taniguchi-Dennis
City of Albany Department of Public Works

Billy Turner
Columbus Water Works

Donna Wies
Union Sanitary District

John Young
American Water

Effective Utility Management Collaborating Organizations

Julia Anastasio
American Public Works Association

John Anderson
American Water Works Association

Peter Cook
National Association of Water Companies

Chris Hornback
National Association of Clean Water Agencies

Jim Horne
*Office of Water
U.S. Environmental Protection Agency*

Eileen O'Neill
Water Environment Federation

Carolyn Peterson
Association of Metropolitan Water Agencies

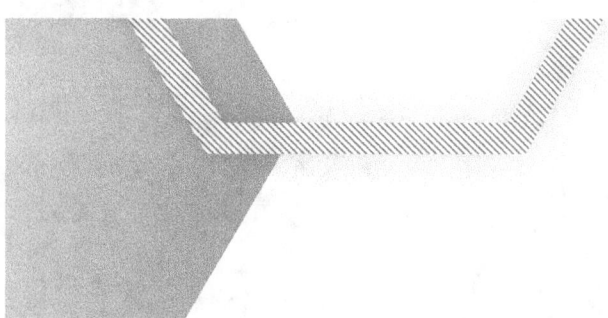

Foreword | Effective Utility Management

Reference herein to any specific commercial products, process, or service by trade name, trademark, manufacturer, or otherwise does not necessarily constitute or imply its endorsement, recommendation, or favoring by the United States Government. The views and opinions of authors expressed herein do not necessarily state or reflect those of the United States Government, and shall not be used for advertising or product endorsement purposes.

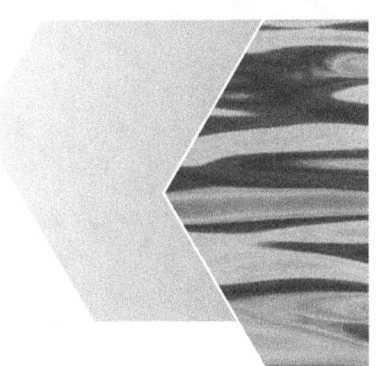

Table of Contents

I.	Effective Utility Management	1
II.	Ten Attributes of Effectively Managed Water Sector Utilities	3
	Ten Attributes of Effectively Managed Water Sector Utilities	4
III.	Keys to Management Success	6
	1. Leadership...	6
	2. Strategic Business Planning.................................	6
	3. Organizational Approaches..................................	7
	4. Measurement...	7
	5. Continual Improvement Management Framework	8
IV.	Where to Begin	10
	Step 1: Assess Current Conditions................................	10
	Step 2: Rank Importance of Attributes.............................	11
	Step 3: Graph Results	13
	Step 4: Choose Attributes.......................................	14
	Step 5: Develop and Implement an Improvement Plan	15
V.	Utility Measures...	16
	Approaching Measurement.......................................	16
	Attribute-Related Measures......................................	17
	List of Attribute-Related Utility Measures	18
VI.	Utility Management Resources	19
VII.	For More Information	20
VIII.	Appendix A: Definitions	21
IX.	Appendix B: Self Assessment...................................	23
	Step 1: Assess Current Conditions................................	23
	Step 2: Rank Importance of Attributes.............................	23
	Step 3: Graph Results	24
X.	Appendix C: Attribute-Related Water Utility Measures..............	25

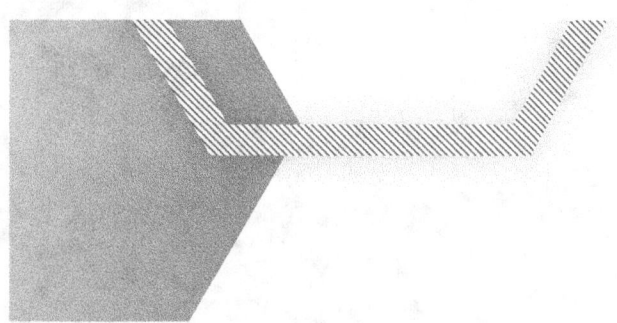

Table of Contents | Effective Utility Management

I. Effective Utility Management

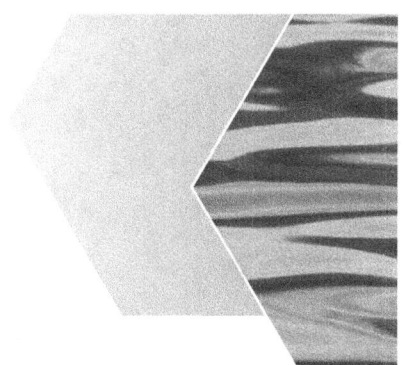

Water and wastewater utilities across the country face common challenges. These include rising costs, aging infrastructure, increasingly stringent regulatory requirements, population changes, and a rapidly changing workforce. While many utility managers find themselves turning from one urgent priority to the next, others have systematically applied effective utility management approaches that have helped them improve their products and services, increase community support, and ensure a strong and viable utility long into the future.

Effective utility management can help water and wastewater utilities enhance the stewardship of their infrastructure, improve performance in many critical areas, and respond to current and future challenges. Addressing these challenges also requires ongoing collaboration between government, industry, elected officials, and other stakeholders.

In May, 2007, six major water and wastewater associations and the U.S. Environmental Protection Agency (EPA) signed an historic agreement pledging to support effective utility management collectively and individually throughout the water sector and to develop a joint strategy to identify, encourage, and recognize excellence in water and wastewater utility management. This Effective Utility Management Primer (Primer) is the result of the agreement among the following organizations:

- Association of Metropolitan Water Agencies (AMWA)
- American Public Works Association (APWA)
- American Water Works Association (AWWA)
- National Association of Clean Water Agencies (NACWA)
- National Association of Water Companies (NAWC)
- United States Environmental Protection Agency (EPA)
- Water Environment Federation (WEF)

Effective utility management is essential to sustaining our nation's water and wastewater infrastructure.

This Primer is designed to help water and wastewater utility managers make practical, systematic changes to achieve excellence in utility performance. It was produced by water and wastewater utility leaders who are committed to helping utility managers improve water and wastewater management. The Primer distills the expertise and experience of these utility leaders into a framework intended to help a utility manager identify and address their most pressing needs through a customized, incremental approach that is relevant to the day-to-day challenges utilities face.

Rather than focusing on just financial or operational goals, this Primer considers all significant aspects of water and wastewater utility management. The Primer has three primary components:

- *The Ten Attributes of Effectively Managed Water Sector Utilities (Attributes)*. These Attributes provide a clear set of reference points and are intended to help utilities maintain a balanced focus on all important operational areas rather than quickly moving from one problem to the next (Section II).
- *Keys to Management Success*. These proven approaches help utilities maximize their resources and improve performance (Section III).
- *Where to Begin–A Self-Assessment Tool*. A utility-tailored self assessment tool helps utility managers identify where to begin improvement efforts. By assessing how a utility performs relative to the Attributes, utility managers can gain a more balanced and comprehensive picture of their organization (Section IV).

Effective utility management is applicable to all utilities, regardless of size or circumstance

In addition, the Primer provides a set of sample measures to help utility managers gauge performance and assess improvement progress (Section V). It also provides links to a web-based "resource toolbox" which offers additional information and guidance on effective utility management (Section VI).

Utility managers and stakeholders can use this Primer in a variety of ways. At one end of the spectrum, the Primer can educate utility staff and stakeholders regarding the range of responsibilities faced by water and wastewater managers. At the other end of the spectrum, it can provide a framework for a utility's long-term strategic planning efforts. Regardless of where a utility is in the spectrum, this Primer can help integrate the Attributes of effective utility management with existing strategic, business, and/or asset management plans.

All water and wastewater utilities can benefit from applying this Primer. Each utility has unique management opportunities and challenges, and this Primer provides guidelines and tools that are relevant to any utility, regardless of size, budget, or circumstance. This Primer's aim is to support all water and wastewater utilities in their common mission of being successful 21st century service providers.

II. Ten Attributes of Effectively Managed Water Sector Utilities

The Ten Attributes of Effectively Managed Water Sector Utilities provide useful and concise reference points for utility managers seeking to improve organization-wide performance. The Attributes describe desired outcomes that are applicable to all water and wastewater utilities. They comprise a comprehensive framework related to operations, infrastructure, customer satisfaction, community welfare, natural resource stewardship, and financial performance.

Water and wastewater utilities can use the Attributes to select priorities for improvement, based on each organization's strategic objectives and the needs of the community it serves. The Attributes are not presented in a particular order, but rather can be viewed as a set of opportunities for improving utility management and operations. Section IV (Where to Begin), provides a basic self-assessment tool to help utilities easily identify needs and opportunities. However, utilities will be able to deliver increasingly efficient, high-quality service by addressing more, and eventually all, of the Attributes. Section V provides several sample performance measures for each of the Attributes.

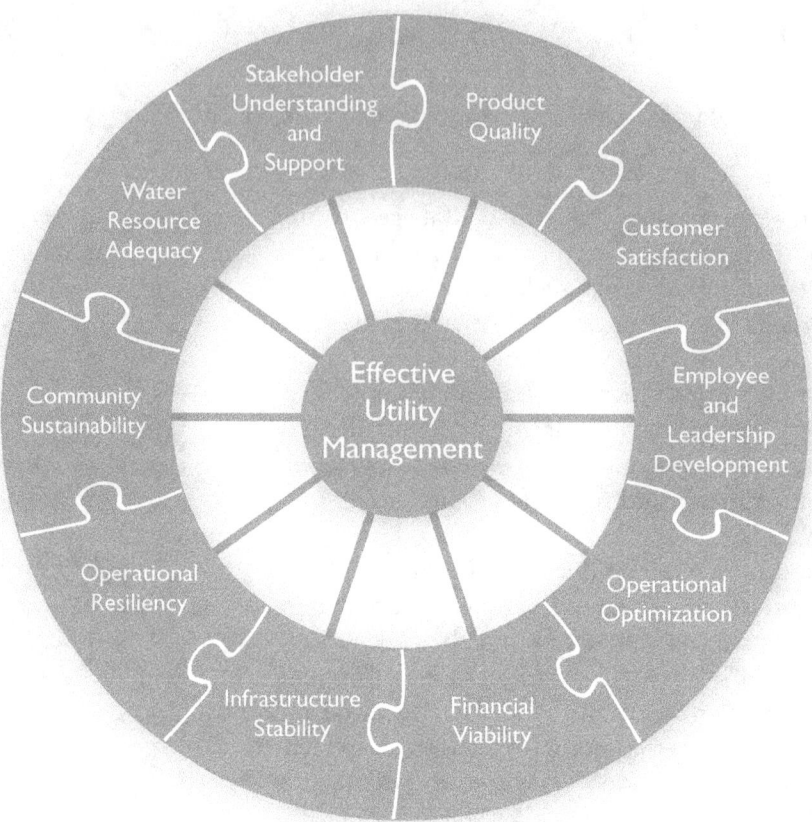

Ten Attributes of Effectively Managed Water Sector Utilities

Ten Attributes of Effectively Managed Water Sector Utilities

Product Quality
Produces potable water, treated effluent, and process residuals in full compliance with regulatory and reliability requirements and consistent with customer, public health, and ecological needs.

Customer Satisfaction
Provides reliable, responsive, and affordable services in line with explicit, customer-accepted service levels. Receives timely customer feedback to maintain responsiveness to customer needs and emergencies.

Employee and Leadership Development
Recruits and retains a workforce that is competent, motivated, adaptive, and safe-working. Establishes a participatory, collaborative organization dedicated to continual learning and improvement. Ensures employee institutional knowledge is retained and improved upon over time. Provides a focus on and emphasizes opportunities for professional and leadership development and strives to create an integrated and well-coordinated senior leadership team.

Operational Optimization
Ensures ongoing, timely, cost-effective, reliable, and sustainable performance improvements in all facets of its operations. Minimizes resource use, loss, and impacts from day-to-day operations. Maintains awareness of information and operational technology developments to anticipate and support timely adoption of improvements.

Financial Viability
Understands the full life-cycle cost of the utility and establishes and maintains an effective balance between long-term debt, asset values, operations and maintenance expenditures, and operating revenues. Establishes predictable rates—consistent with community expectations and acceptability—adequate to recover costs, provide for reserves, maintain support from bond rating agencies, and plan and invest for future needs.

Infrastructure Stability
Understands the condition of and costs associated with critical infrastructure assets. Maintains and enhances the condition of all assets over the long-term at the lowest possible life-cycle cost and acceptable risk consistent with customer, community, and regulator-supported service levels, and consistent with anticipated growth and system reliability goals. Assures asset repair, rehabilitation, and replacement efforts are coordinated within the community to minimize disruptions and other negative consequences.

Operational Resiliency
Ensures utility leadership and staff work together to anticipate and avoid problems. Proactively identifies, assesses, establishes tolerance levels for, and effectively manages a full range of business risks (including legal, regulatory, financial, environmental, safety, security, and natural disaster-related) in a proactive way consistent with industry trends and system reliability goals.

Community Sustainability
Is explicitly cognizant of and attentive to the impacts its decisions have on current and long-term future community and watershed health and welfare. Manages operations, infrastructure, and investments to protect, restore, and enhance the natural environment; efficiently uses water and energy resources; promotes economic vitality; and engenders overall community improvement. Explicitly considers a variety of pollution prevention, watershed, and source water protection approaches as part of an overall strategy to maintain and enhance ecological and community sustainability.

Water Resource Adequacy
Ensures water availability consistent with current and future customer needs through long-term resource supply and demand analysis, conservation, and public education. Explicitly considers its role in water availability and manages operations to provide for long-term aquifer and surface water sustainability and replenishment.

Stakeholder Understanding and Support
Engenders understanding and support from oversight bodies, community and watershed interests, and regulatory bodies for service levels, rate structures, operating budgets, capital improvement programs, and risk management decisions. Actively involves stakeholders in the decisions that will affect them.

III. Keys to Management Success

The Keys to Management Success are comprised of frequently used management approaches and systems that experience indicates help water and wastewater utilities manage more effectively. They create a supportive climate for a utility as it works towards the outcomes outlined in the Attributes, and they can help integrate the utility's improvement efforts across the Attributes. The Keys to Management Success are listed below.

1. Leadership

Effective leadership produces organizational alignment and clear direction

Leadership is critical to effective utility management, particularly in the context of driving and inspiring change within an organization. "Leadership" refers both to individuals who can be effective champions for improvement, and to teams that provide resilient, day-to-day management continuity and direction. Effective leadership ensures that the utility's direction is understood, embraced, and followed on an ongoing basis throughout the management cycle. Leadership has an important responsibility to communicate with the utility's stakeholders and customers. It further reflects a commitment to organizational excellence, leading by example to establish and reinforce an organizational culture that embraces positive change and strives for continual improvement. Organizational improvement efforts require commitment from the utility's leadership.

2. Strategic Business Planning

Strategic business planning is an important tool for achieving balance and cohesion across the Attributes. A strategic plan provides a framework for decision making by:

- Assessing current conditions, strengths and weaknesses;
- Assessing underlying causes and effects; and
- Establishing vision, objectives, and strategies.

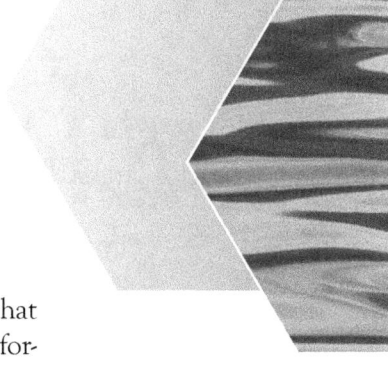

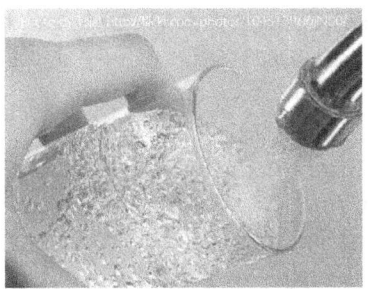
It establishes specific implementation steps that will move a utility from its current level of performance to achieving its vision.

Preparation of a strategic business plan involves taking a long-term view of utility goals and operations and establishing a clear vision and mission. When developed, the strategic business plan will drive and guide utility objectives, measurement efforts, investments, and operations. A strategic plan can help explain the utility's conditions, goals, and plans to staff and stakeholders, stimulate change, and increase engagement in improvement efforts.

After developing a strategic business plan, it is important that the utility integrates tracking of progress into its management framework.

3. Organizational Approaches

There are a variety of organizational approaches that contribute to overall effective utility management and that are critical to the success of management improvement efforts. These include:

- Actively engaging employees in improvement efforts (helping to identify improvement opportunities, participating in cross-functional improvement teams, etc.);
- Deploying an explicit change management process that anticipates and plans for change and encourages staff at all levels to embrace change; and
- Utilizing implementation strategies that seek, identify, and celebrate early, step-by-step victories.

4. Measurement

Measurement is critical to management improvement efforts associated with the Attributes and is the backbone of successful continual improvement management and strategic business planning. A measurement system serves many vital purposes, including focusing attention on key issues, clarifying expectations, facilitating decision making, and, most importantly, learning and improving. As one utility manager put it, "You can't improve what you don't measure." Successful measurement efforts often are:

> *"You can't improve what you don't measure."*

- Viewed as a continuum starting with basic internal tracking, and, as needed and appropriate, moving to more sophisticated baselining and trend analysis, development of key performance indicators, and inclusion of externally oriented measures which address community sustainability interests;
- Driven by and focused on answering questions critical to effective internal management and external stakeholder needs (e.g., information needed to allow governing bodies to comfortably support large capital investments); and
- Supported by a well-defined decision framework assuring results are evaluated, communicated, and responded to in a timely manner.

Deciding where to start and what to measure can be challenging. Measures can also be taken out of context. Therefore, while an essential tool in the self-improvement process, measurement is not the only tool and should be approached, structured, and used thoughtfully. Section V includes sample performance measures that can be used in conjunction with utility-specific baselines and targets.

5. Continual Improvement Management Framework

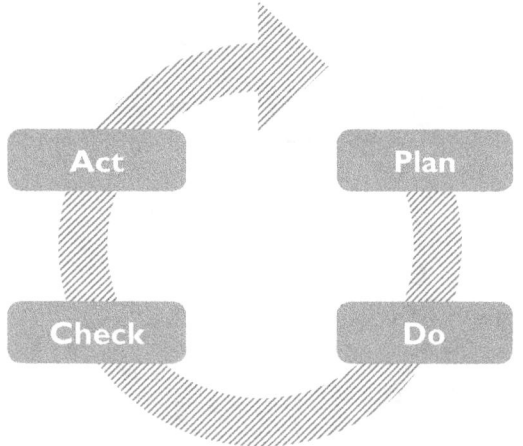

A continual improvement management framework is usually implemented through a complete, start-to-finish management system, frequently referred to as a "Plan-Do-Check-Act" framework. This framework plays a central role in effective utility management and is critical to making progress on the Attributes. Continual improvement management includes:

- Conducting an honest and comprehensive self-assessment to identify management strengths, areas for improvement, priority needs, etc.;
- Conducting frequent sessions among interested parties to identify improvement opportunities;
- Following up on improvement projects underway;
- Establishing and implementing performance measures and specific internal targets associated with those measures;
- Defining and implementing related operational requirements, practices, and procedures;
- Establishing supporting roles and responsibilities;
- Implementing measurement activities such as regular evaluation through operational and procedural audits; and
- Responding to evaluations through the use of an explicit change management process.

This "Plan-Do-Check-Act" continual improvement framework is quite effective when applied internally. It can also be enhanced by using gap analysis, establishment of standard operating procedures, internal trend analysis and external benchmarking, best practice review, and other continual improvement tools. The framework can help utilities understand improvement opportunities and establish explicit service levels, guide investment and operational decisions, form the basis for ongoing measurement, and provide the ability to communicate clearly with customers and key stakeholders.

The Resource Toolbox described in Section VI, Utility Management Resources, provides links to resources that support utilization of the Keys to Management Success.

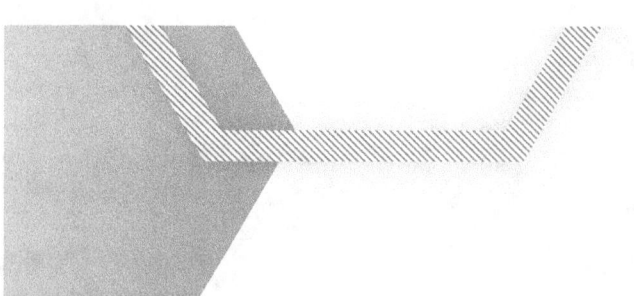

IV. Where to Begin

Step 1
Candidly Assess Current Conditions

Step 2
Rank Importance of Each Attribute to Your Utility

Step 3
Graph Attributes to Determine Importance and Level of Achievement

Step 4
Choose Attributes

Step 5
Develop and Implement an Improvement Plan

There are many ways to improve utility performance and each utility is unique. Many utilities may choose to start small and make improvements step by step, perhaps by working on projects that will yield early successes. Other utilities may choose to take on several ambitious change efforts simultaneously. Some may prefer to enhance their strengths, while others will prefer to focus on addressing weaknesses. Each utility should determine for itself the most important issue to address, based on its own strategic objectives, priorities, and the needs of the community it serves.

A candid assessment of current performance is often a useful first step in identifying options for improvement. It also establishes a quantifiable baseline from which to measure progress. As conditions change, future reassessments will reveal new opportunities and new priorities.

The following self assessment tool can help water and wastewater managers evaluate their utility's current performance against internal goals or specific needs and determine where to focus improvement efforts. It can be completed by an individual manager, but would also be useful as a vehicle for conversation and consensus building among the utility's management team and other appropriate stakeholders, such as oversight bodies, community and watershed interests, and regulatory authorities.

The assessment tool has five steps: 1) Assess current conditions; 2) Rank the importance of each Attribute for your utility; 3) Chart the results; 4) Choose one or more Attributes to focus on; and 5) Develop and implement an improvement plan.

The Self Assessment can also be found in Appendix B.

Step 1: Assess Current Conditions

On a 1-to-5 scale, assess current conditions by rating your utility's systems and approaches and current level of achievement for each Attribute. Consider the degree to which your current management systems effectively support each of the Attributes and their component parts. Consider all components of each Attribute and gauge your rating accordingly. Use these descriptions to guide your rating.

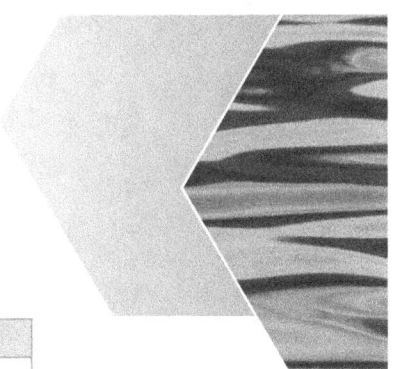

Rating	Description
1.	Effective, systematic approach and implementation; consistently achieve goals.
2.	Workable systems in place; mostly achieve goals.
3.	Partial systems in place with moderate achievement, but could improve.
4.	Occasionally address this when specific need arises.
5.	No system for addressing this.

Step 2: Rank Importance of Attributes

Rank the importance of each Attribute to your utility, based on your utility's vision, goals, and specific needs. The ranking should reflect the interests and considerations of all stakeholders (managers, staff, customers, regulators, elected officials, community and watershed interests, shareholders, and others).

There are ten Attributes; considering long-term importance to your utility, rank the most important Attribute 1, the second most important 2, and so on. The least important Attribute would be ranked 10. Your ranking of each Attribute's importance might be influenced by current or expected challenges in that particular area, recent accomplishments in addressing these issues, or other factors. Importance ranking is likely to change over time as internal and external conditions change.

As you fill in numbers on the table below, please note that your analysis for Step 1 (rating achievement) should be separate and independent from your analysis for Step 2 (ranking importance).

Attribute	Attribute Components	Step 1: Rate Achievement (1-5)	Step 2: Rank Importance (1-10)
Product Quality (PQ)	○ Complies with regulatory and reliability requirements. ○ Consistent with customer, public health, and ecological needs.		
Customer Satisfaction (CS)	○ Provides reliable, responsive, and affordable services. ○ Receives timely customer feedback. ○ Responsive to customer needs and emergencies.		

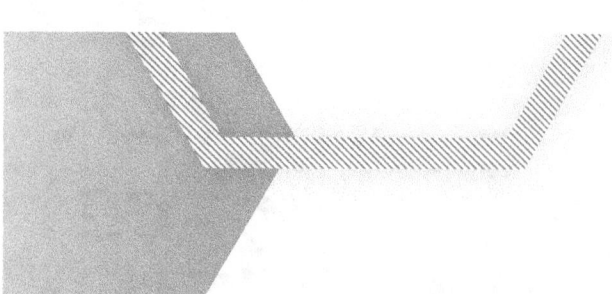

Rating and Ranking Table, *continued*

Attribute	Attribute Components	Step 1: Rate Achievement (1-5)	Step 2: Rank Importance (1-10)
Employee and Leadership Development (ED)	○ Recruits and retains competent workforce. ○ Collaborative organization dedicated to continual learning and improvement. ○ Employee institutional knowledge retained and improved. ○ Opportunities for professional and leadership development. ○ Integrated and well-coordinated senior leadership team.		
Operational Optimization (OO)	○ Ongoing performance improvements. ○ Minimizes resource use and loss from day-to-day operations. ○ Awareness and timely adoption of operational and technology improvements.		
Financial Viability (FV)	○ Understands full life-cycle cost of utility. ○ Effective balance between long-term debt, asset values, operations and maintenance expenditures, and operating revenues. ○ Predictable and adequate rates.		
Infrastructure Stability (IS)	○ Understands the condition of and costs associated with critical infrastructure assets. ○ Maintains and enhances assets over the long-term at the lowest possible life-cycle cost and acceptable risk. ○ Repair efforts are coordinated within the community to minimize disruptions.		
Operational Resiliency (OR)	○ Staff work together to anticipate and avoid problems. ○ Proactively establishes tolerance levels and effectively manages risks (including legal, regulatory, financial, environmental, safety, security, and natural disaster-related).		

Rating and Ranking Table, *continued*

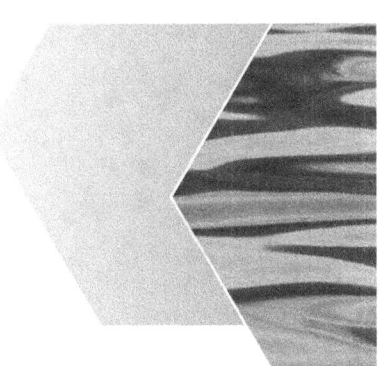

Attribute	Attribute Components	Step 1: Rate Achievement (1-5)	Step 2: Rank Importance (1-10)
Community Sustainability (SU)	○ Attentive to impacts on community and watershed health and welfare. ○ Operations enhance natural environment. ○ Efficiently use water and energy resources; promote economic vitality; and engender overall community improvement. ○ Maintain and enhance ecological and community sustainability including pollution prevention, watershed, and source water protection.		
Water Resource Adequacy (WA)	○ Ensures water availability through long-term resource supply and demand analysis, conservation, and public education. ○ Manages operations to provide for long-term aquifer and surface water sustainability and replenishment.		
Stakeholder Understanding and Support (SS)	○ Engenders understanding and support from oversight bodies, community and watershed interests, and regulatory bodies for service levels, rate structures, operating budgets, capital improvement programs, and risk management decisions. ○ Actively involves stakeholders in the decisions that will affect them.		

Step 3: Graph Results

Graph each Attribute based on your rating and ranking. For example, if you rated Product Quality (PQ) 4 for achievement and ranked it 3 for importance, you would place it on the graph as illustrated below. Similarly, if you rated Customer Satisfaction (CS) 3 for achievement and ranked it 5 for importance, you would place it on the graph as illustrated below. A blank graph is provided in Appendix B.

		5									
	Lower Achievement	4			PQ						
Rating		3					CS				
	Higher Achievement	2									
		1									
			1	2	3	4	5	6	7	8	9	10
				More Important					Less Important		
			Ranking								

Step 4: Choose Attributes

The goal of effective utility management is to establish high-achieving systems and approaches for each Attribute. Ultimately, utilities should strive to improve performance for all Attributes until each can be charted in the lower half of the table (high achieving). Utility managers may wish to focus on one or a few Attributes at a time, aiming to eventually ensure that all Attributes have been addressed and improved upon over time.

Examining the results of the charting exercise in Step 3 can help identify Attributes to focus on. Attributes that graph into the *blue quadrant* are both very important (ranked 1-5), and under-developed (rated 3-5). These Attributes are strong candidates for improvement efforts. Attributes that fall in the lower left-hand quadrant are both important and well-developed. Some utilities may choose to focus on these areas to continue further improving upon important and well-developed areas, due to their long-term importance (for example, water resource adequacy). Specifically examining these areas may also help a utility identify success factors which would be helpful in addressing areas needing improvement. Others may choose to focus on Attributes that would lead to early successes to build confidence in effecting change, Attributes that maximize benefit relative to the utility's key goals, or Attributes that minimize risks (e.g., fines, penalties, lawsuits, poor public perception).

The choice to embark on improvements in one or more areas is up to the judgment of utility managers, and may also involve consideration of resources (staff and financial), leadership support, and other competing activities. Applying strategic business planning, measurement, and other Keys to Management Success is very important for moving each Attribute over time to the "well-developed" quadrants.

Step 5: Develop and Implement an Improvement Plan

Once you choose to improve one or more Attributes, the next step is to develop and implement a plan for making the desired improvements. Effective improvement plans commonly include the following features:

- A "gap" analysis to identify root causes of under-performance. This analysis would describe the utility's performance goals, its current position relative to its goals, and the reasons for not achieving its goals;
- Development of a utility-specific plan and/or strategy to achieve performance goals and address the root causes. The plan should consider how to incorporate customer and, as appropriate, broader stakeholder interests;
- Specific tasks, tactics, or management adjustments necessary to implement the utility's strategy;
- Utility-specific measures to track progress toward achievement of performance goals; and
- A timeframe for follow-up measurement to assess the degree of accomplishment and potential need for additional effort.

Utilities may also find it useful to appoint an overall improvement program manager to oversee individual improvement projects.

The improvement plan should be developed and implemented within the context of strategic business planning, the "Plan-Do-Check-Act" continual improvement framework, and other components of the Keys to Management Success discussed in Section III.

V. Utility Measures

Measuring performance is one of the keys to utility management success. This section of the Primer provides ideas about how to approach measurement and then offers measures for each Attribute to help understand a utility's status and progress.

Approaching Measurement

There are two general approaches to performance measurement. Internal performance measurement, which is the focus of this Primer, involves evaluating current internal utility performance status and trends. It can also include comparison of outcomes or outputs relative to goals, objectives, baseline status, targets, and standards. Benchmarking—*which is not this Primer's focus*—is the overt comparison of similar measures or processes across organizations to identify best practices, set improvement targets, and measure progress within or sometimes across sectors. A utility may decide to engage in benchmarking for its own internal purposes or in a coordinated fashion with others.

While performance measures should be tailored to the specific needs of your utility, the following guidelines can help you identify useful measures and apply them effectively.

1. Select measures that support the organization's strategic objectives, mission, and vision, as well as the ten Attributes.
2. Select the right number, level, and type of measures for your organization. Consider how measures can be integrated as a cohesive group (e.g., start with a small set of measures across broad categories and increase number and specificity over time as needed), and consider measures that can be used by different audiences within the organization.
3. Measuring performance will not necessarily require additional staff, but will require resources. Allocate adequate resources to get the effort off to a good start, and fine tune over time to balance the level of measurement effort with the benefit to the organization.
4. Develop clear, consistent definitions for each measure. Identify who is responsible for collecting the data, and how the data will be tracked and reported.
5. Engage the organization at all levels in developing, tracking, and reporting measures, but also assign someone in the organization the role of championing and coordinating the effort.

6. Set targets rationally, based on criteria such as customer expectations, improvement over previous years, industry performance, or other appropriate comparisons. Tie targets to improving performance in the Attributes.
7. Select and use measures in a positive way to improve decision making, clarify expectations, and focus attention, not just to monitor, report, and control.
8. When selecting measures, consider how they relate to one another. Look for cause-and-effect relationships; for example, how improvements in product quality could result in increased customer satisfaction.
9. Develop an effective process to evaluate and respond to results. Identify how, when, and to whom you will communicate results.
10. Incorporate the "Plan-Do-Check-Act" cycle approach into evaluating both the specific measures and the system as a whole. Regularly review the performance measurement system for opportunities to improve.

... and remember to celebrate your measured and documented successes!

Attribute-Related Measures

The list below provides a limited list of targeted, Attribute-related measures. Taken as a whole, the measures provide a utility with a cohesive, approachable, and generally applicable starting place for gauging progress relative to the Ten Attributes. The list, for brevity, contains measure "headlines" for each Attribute; Appendix C provides further explanation and, where applicable, example calculations.

You can choose and tailor the measures to your own needs and unique, local circumstances. They are intended for your own internal use, even as certain measures (e.g., those noted as QualServe Indicators) can support benchmarking purposes. In these cases, the measures have been selected because they are relevant to the Attributes, have been tested and are in use by utilities, are supported by reference information useful for implementation, and generally can act as a good starting point for Attribute-related progress assessment.

As described in Appendix C, the measures are both quantitative and qualitative. Most are quantitative and include generally applicable example calculations. The qualitative "measures" encourage active assessment of the management area and most have a "yes/no" format.

Like the Attributes themselves, certain measures focus on core utility operations. Several measures reflect emerging utility issues, challenges, or opportunities that have received increasing attention from a growing number of utility managers. Other measures may reflect broader interests that are worthy of consideration from a broader community perspective.

List of Attribute-Related Utility Measures

See Appendix C for measure descriptions and details.

Product Quality
1. Product quality regulatory compliance
2. Product quality service delivery

Customer Satisfaction
1. Customer complaints
2. Customer service delivery
3. Customer satisfaction

Employee and Leadership Development
1. Employee retention and satisfaction
2. Management of core competencies
3. Workforce succession preparedness

Operational Optimization
1. Resource optimization
2. Water management efficiency

Financial Viability
1. Budget management effectiveness
2. Financial procedure integrity
3. Bond ratings
4. Rate adequacy

Infrastructure Stability
1. Asset inventory
2. Asset (system) renewal/replacement
3. Water distribution/collection system integrity
4. Planned maintenance

Operational Resiliency
1. Recordable incidents of injury or illnesses
2. Insurance claims
3. Risk assessment and response preparedness
4. Ongoing operational resiliency
5. Operational resiliency under emergency conditions

Community Sustainability
1. Watershed-based infrastructure planning
2. Green infrastructure
3. Greenhouse gas emissions
4. Service affordability

Water Resource Adequacy
1. Water supply adequacy
2. Supply and demand management

Stakeholder Understanding and Support
1. Stakeholder consultation
2. Stakeholder satisfaction
3. Internal benefits from stakeholder input
4. Comparative rate rank
5. Media/press coverage

VI. Utility Management Resources

As a companion resource to this Primer, the Collaborating Organizations developed an online Resource Toolbox which offers additional information and guidance on effective utility management. The Toolbox provides a compilation of resources from the seven Collaborating Organizations designed to help the water and wastewater utility community further improve the management of its infrastructure.

The Resource Toolbox is organized according to the Ten Attributes of Effectively Managed Water Sector Utilities and five Keys to Management Success, providing a set of resources relevant to each Attribute and Key. The Toolbox also includes information on where to find these resources.

The Resource Toolbox is located at the website for the Effective Utility Management initiative, at www.watereum.org.

VII. For More Information

This Primer was developed through a collaborative partnership with the following groups. More information about this partnership can be found on their websites or by contacting specific individuals directly.

American Public Works Association
Julia Anastasio
Senior Manager of Government Affairs
1401 K Street, NW, 11th Floor
Washington DC 20005
janastasio@apwa.net
202.218.6750
www.apwa.net

American Water Works Association
Ed Baruth
Director, Volunteer and Technical Support Group
6666 W. Quincy Ave.
Denver CO 80235
ebaruth@awwa.org
303.347.6176
www.awwa.org

Association of Metropolitan Water Agencies
Carolyn Peterson
Director of Communications and Public Affairs
1620 I Street, NW
Washington DC 20006
peterson@amwa.net
202.331.2820
www.amwa.net

National Association of Clean Water Agencies
Chris Hornback
Senior Director, Regulatory Affairs
1816 Jefferson Place, NW
Washington DC 20036
chornback@nacwa.org
202.833.9106
www.nacwa.org

National Association of Water Companies
Peter Cook
Executive Director
2001 L Street, NW, Suite 850
Washington DC 20036
peter@nawc.com
202.833.2100
www.nawc.org

U.S. Environmental Protection Agency
Jim Horne
US EPA, Office of Wastewater Management
1200 Pennsylvania Avenue, NW
Room 7111 – EPA East
Washington DC 20460
horne.james@epa.gov
202.564.0571
www.epa.gov/waterinfrastructure

Water Environment Federation
Eileen O'Neill
Chief Technical Officer
601 Wythe Street
Alexandria VA 22314
eoneill@wef.org
703.684.2462
www.wef.org/ScienceTechnologyResources/UtilityManagement

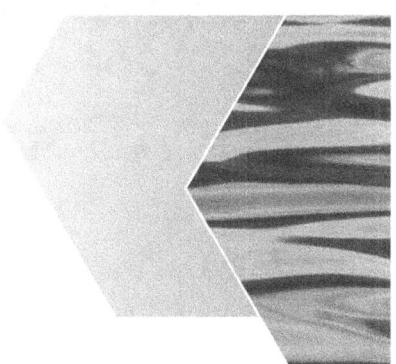

VIII. Appendix A: Definitions

The following terms are presented in this Primer. These definitions provide a brief overview of their meaning.

- **Attribute**: A characteristic or outcome of a utility that indicates effective performance.
- **Benchmarking**: The comparison of similar processes or measures across organizations and/or sectors to identify best practices, set improvement targets, and measure progress.
- **Effective Utility Management**: Management that improves products and services, increases community support, and ensures a strong and viable utility into the future.
- **Gap analysis**: Defining the present state of an enterprise's operations, the desired or "target" state, and the gap between them.
- **Internal trend analysis**: Comparison of outcomes or outputs relative to goals, objectives, baselines, targets, and standards.
- **Life-cycle cost**: The total of all internal and external costs associated with a product, process, or activity throughout its entire life cycle – from raw materials acquisition to manufacture/construction/installation, operation and maintenance, recycling, and final disposal.
- **Performance measurement**: Evaluation of current status and trends; can also include comparison of outcomes or outputs relative to goals, objectives, baselines, targets, standards, other organizations' performance or processes (typically called benchmarking), etc.
- **Operations and maintenance expenditure**: Expenses used for day-to-day operation and maintenance of a facility.
- **Operating revenue**: Revenue realized from the day-to-day operations of a utility.
- **Performance measure**: A particular value or characteristic designated to measure input, output, outcome, efficiency, or effectiveness.
- **Source water protection**: Efforts to prevent water quality degradation in streams, rivers, lakes, or underground aquifers used as public drinking water supplies.
- **Standard operating procedure**: A prescribed procedure to be followed routinely; a set of instructions having the force of a directive, covering those features of operations that lend themselves to a definite or standardized procedure without loss of effectiveness.

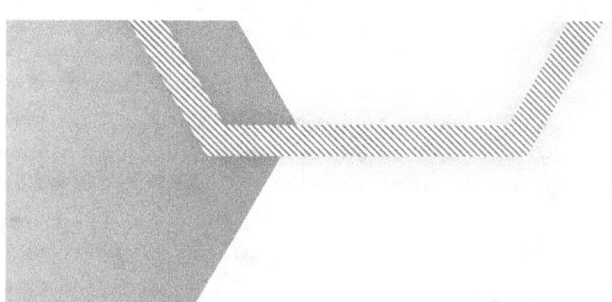

- **Strategic plan**: An organization's process of defining its goals and strategy for achieving those goals. Often entails identifying an organization's vision, goals, objectives, and targets over a multi-year period of time, as well as setting priorities and making decisions on allocating resources, including capital and people, to pursue the identified strategy.
- **Stewardship**: The careful and responsible management of something entrusted to a designated person or entity's care; the responsibility to properly utilize its resources, including its people, property, and financial and natural assets.
- **Sustainability**: The use of natural, community, and utility resources in a manner that satisfies current needs without compromising future needs or options.
- **Watershed health**: The ability of ecosystems to provide the functions needed by plants, wildlife, and humans, including the quality and quantity of land and aquatic resources.

IX. Appendix B: Self Assessment

Step 1: Assess Current Conditions

On a 1-to-5 scale, assess current conditions by rating your utility's systems and approaches and current level of achievement for each Attribute. Consider the degree to which your current management systems effectively support each of the Attributes and their component parts. Consider all components of each Attribute and gauge your rating accordingly. Use these descriptions to guide your rating.

Rating	Description
1.	Effective, systematic approach and implementation; consistently achieve goals.
2.	Workable systems in place; mostly achieve goals.
3.	Partial systems in place with moderate achievement, but could improve.
4.	Occasionally address this when specific need arises.
5.	No system for addressing this.

Mark your answers in the Step 1 column of the table on the next page.

Step 2: Rank Importance of Attributes

Rank the importance of each Attribute to your utility, based on your utility's vision, goals, and specific needs. The ranking should reflect the interests and considerations of all stakeholders (managers, staff, customers, regulators, elected officials, community and watershed interests, shareholders, and others).

There are ten Attributes; considering long-term importance to your utility, rank the most important Attribute 1, the second most important 2, and so on. The least important Attribute would be ranked 10. Your ranking of each Attribute's importance might be influenced by current or foreseeable challenges in that particular area, recent accomplishments in addressing these issues, or other factors. Importance ranking is likely to change over time as internal and external conditions change.

Mark your answers in the Step 2 column of the table on the next page. As you fill in numbers, please note that your analysis for Step 1 (rating achievement) should be separate and independent from your analysis for Step 2 (ranking importance).

Attribute	Step 1: Rate Achievement (1-5)	Step 2: Rank Importance (1-10)
Product Quality (PQ)		
Customer Satisfaction (CS)		
Employee and Leadership Development (ED)		
Operational Optimization (OO)		
Financial Viability (FV)		
Infrastructure Stability (IS)		
Operational Resiliency (OR)		
Community Sustainability (SU)		
Water Resource Adequacy (WA)		
Stakeholder Understanding and Support (SS)		

Step 3: Graph Results

Graph each Attribute based on your rating and ranking.

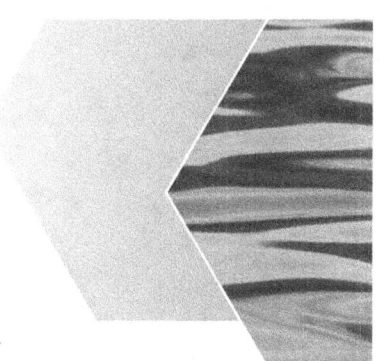

X. Appendix C: Attribute-Related Water Utility Measures

This Appendix provides more detailed information on the measures offered in Section V of the Primer, including descriptions and example calculations and questions.

Product Quality

1. Product quality regulatory compliance

Description: Water product quality compliance, particularly with regards to 40 CFR Part 141 (the National Primary Drinking Water Regulations), the National Pollutant Discharge Elimination System, and any other relevant federal (Clean Water Act, Safe Drinking Water Act, etc.) or state statute/regulations and permit requirements. The scope can include the quality of all related products, including drinking water, fire suppression water, treated effluent, reused water, and biosolids, as well as quality-related operating requirements such as pressure and number of sewer overflows.

Example calculations:

- *Drinking water compliance rate* (percent): 100 X (number of days in full compliance for the year ÷ 365 days). This is a QualServe Indicator.[1]
- *Wastewater treatment effectiveness rate* (percent): 100 X (365 – total number of standard noncompliance days ÷ 365 days). This is a QualServe Indicator.[2]
- *Number, type, and frequency of "near (compliance) misses"*: For example, reaching 80-95% of allowable levels of "X" during reporting period, typically per month. Tracking this type of measure could be used to improve performance in these "near miss" areas before violations occur.

2. Product quality service delivery

Description: This measure assesses delivery of product quality service based on utility-established objectives and service level targets. It focuses on non-regulatory performance targets.

---

[1] This is one of the 22 Performance Indicators from the Qualserve program, a voluntary quality improvement program designed for water and wastewater utilities by the American Water Works Association and the Water Environment Federation. Reference from the American Water Works Association and the Awwa Research Foundation, *Selection and Definition of Performance Indicators for Water and Wastewater Utilities*, p. 57. 2004. Note: This material is copyrighted and any reprinting must be by permission of the American Water Works Association.
[2] Ibid., p. 71. 2004.

Example calculations:

- *Drinking water flow and pressure* (percent): 100 X [number of customers with less than (flow of "X" gallons per minute (gpm) and pressure of "Y" pounds per square inch (psi)—levels set by utility) ÷ total number of customers] (during reporting period, typically per month).
- *Fire suppression water flow and pressure* (percent): 100 X [hours of time when (flow of "X" gpm and pressure of "Y" psi—levels set by utility) is available for fire suppression at maximum day demand ÷ total number of hours when fire suppression water should be available at maximum day demand] (during reporting period, typically per month).
- *Service interruptions* (percent): 100 X (number of active account customers experiencing a service interruption of greater than 1 hour ÷ total number of customers during reporting period) (typically per month). Note: the utility may elect to measure planned and unplanned interruptions separately.
- *Water quality goals met/not met*: Number of days in reporting period (typically one month) where utility-defined beyond-compliance targets are met/not met.
- *Sewer backups* (if not included in permit requirements) (amount and percent): Number of customers experiencing backups each year; 100 X (number of customers experiencing backups each year ÷ total number of customers).
- *Sewer overflows* (if not included in permit requirements): Number of sewer overflows per 100 miles of collection system piping.
- *Water reuse* (amount and percent):
  - Amount: Amount of water supplied that is from reused/recycled sources.
  - Percent: 100 X (amount of water supplied that is from reused/recycled water ÷ total amount of water supplied).

  Then, as desired, these amounts can be broken into recipients/applications (e.g., irrigation, agriculture, industrial processes, etc.).
- *Biosolids put to beneficial use* (percent): 100 X (amount of biosolids produced that are put to a beneficial use ÷ total amount of biosolids produced) (in wet tons per year).

Customer Satisfaction

1. Customer complaints

Description: This measure assesses the complaint rates experienced by the utility, with individual quantification of customer service and core utility service complaints.[3] As a "passive measure," it will not likely be numerically representative (i.e., a statistically valid customer sample group) and is a "starting point" measure for understanding customer service problems.

Example calculations:

- Number of complaints per 1,000 customers per reporting period, recorded as either customer service or technical quality complaints. These calculations are based on the QualServe Customer Service Complaints/Technical Quality Complaints Indicator.
    - *Customer service complaint rate*: 1,000 X (customer service associated complaints ÷ number of active customer accounts). This is a QualServe Indicator.[4]
    - *Technical quality complaint rate*: 1,000 X (technical quality associated complaints ÷ number of active customer accounts). This is a QualServe Indicator.[5]

For both calculations, utilities may wish to subcategorize complaints by type and aspect (e.g., customer service into billing, problem responsiveness, interruptions, etc., and technical quality into service deficiencies such as taste, odor, appearance, flow/pressure, etc.) and by type of customer (e.g., residential, industrial, commercial, etc.)

2. Customer service delivery

Description: This measure requires the utility, based on internal objectives and customer input, to set desirable customer service levels, then determine an appropriate (target) percentage of time to meet the performance levels. Once established, the utility can track how often it meets the service levels, helping the utility to determine how well customer needs are being satisfied (e.g., have 95 percent of service calls received a response within 60 minutes). A utility can average across individual measures to determine the overall percentage of service level commitments met.

---

[3] From AWWA and AwwaRF, *Selection and Definition of Performance Indicators for Water and Wastewater Utilities*, p. 41. 2004. Note: This material is copyrighted and any reprinting must be by permission of the American Water Works Association
[4] Ibid., p. 41.
[5] Ibid., p. 42.

Example calculations:

- *Call responsiveness* (percent): 100 X (number of calls responded to within "X" minutes ÷ total number of calls during reporting period) (typically per month).
- *Error-driven billing adjustment rate* (percent): 100 X (number of error-driven billing adjustments during reporting period ÷ number of bills generated during reporting period). This is a QualServe Indicator.[6]
- *Service start/stop responsiveness* (percent): 100 X (number of stop/start service orders processed within "X" days ÷ total number of stop/start service orders during reporting period).
- *First call resolution* (percent): 100 X (number of calls for which problem was resolved/fixed/scheduled to be fixed at the time of the first call ÷ total number of calls during reporting period).

3. Customer satisfaction

Description: This is an overarching customer satisfaction measure based on requested customer feedback (surveys), not calls received or internal customer satisfaction service level commitments. A utility can measure customer satisfaction immediately after service provision or use a periodically performed, more comprehensive customer satisfaction survey. After-service surveys are simpler and easier for the utility to develop and implement without professional advice, but they tend to over represent the most satisfied (e.g., those who just received service) and the most dissatisfied (e.g., those who just called with complaints) customers. Comprehensive surveys can provide statistical validity enabling extrapolation to the population served. A utility can verify survey information through customer conversations, either as follow up to a survey, during public meetings or focus groups, or by some other method (e.g., individual telephone calls).

Example calculation:

- *Overall customer satisfaction*: Percent of positive or negative customer satisfaction survey responses based on a statistically valid survey or on an immediately after-service survey. Satisfaction responses can be divided into categories such as: highly satisfied/satisfied/moderately satisfied/unsatisfactory; exceeding expectations/meeting expectations/not meeting expectations; numerical scales (e.g., 1-5); or other divisions. Customer satisfaction information is often also gathered and assessed by topic areas such as product quality, service reliability, billing accuracy, customer service, costs/rates/value, crew courtesy, notification around street construction/service interruptions, etc.

---

[6] From AWWA and AwwaRF, *Selection and Definition of Performance Indicators for Water and Wastewater Utilities*, p. 49. 2004. Note: This material is copyrighted and any reprinting must be by permission of the American Water Works Association.

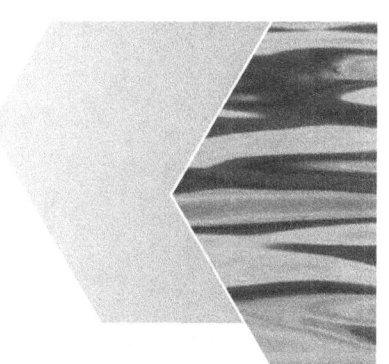

Employee and Leadership Development

1. Employee retention and satisfaction

Description: This measure gauges a utility's progress toward developing and maintaining a competent and stable workforce, including utility leadership.

Example calculations:

- *Employee turnover rate* (percent): 100 X (number of employee departures ÷ total number of authorized positions per year). Can be divided into categories such as:
  - *Voluntary turnover* (percent): 100 X (number of voluntary departures ÷ total number of authorized positions per year). (Perhaps the best indicator of retention problems.)
  - *Retirement turnover* (percent): 100 X (number of retirement departures ÷ authorized positions per year). (Measures loss/retention of institutional knowledge.)
  - *Experience turnover* (percent): 100 X (number of years of experience represented by all departures ÷ total years of experience with the organization) (at the beginning of the year). (These are harder data to collect but provide a good assessment of institutional knowledge loss potential and therefore the need to retain/capture institutional knowledge.)
- *Employee job satisfaction* (percent): 100 X (number of employees with "X" job satisfaction level ÷ total number of employees) (based on implementation and monitoring over time of a comprehensive employee survey). Can be divided into work type or job classification categories, etc., and cover overall satisfaction and topics deemed relevant to longer-term employee satisfaction and retention, such as:
  - Compensation and benefits
  - Management
  - Professional development and long-term advancement opportunities
  - Work and teamwork
  - Procedures
  - Fairness and respect
  - Communication

2. Management of core competencies

Description: This measure assesses the utility's investment in and progress toward strengthening and maintaining employee core competencies.

Example calculations and assessment areas:

- *Presence of job descriptions and performance expectations*: Does your organization have and maintain current job descriptions and related performance expectations (yes/no)?
- *Training hours per employee*: Total of qualified formal training hours for all employees ÷ total FTEs worked by employees during the reporting period. This is a QualServe Indicator.[7]
- *Certification coverage* (percent): 100 X (number of certifications achieved or maintained ÷ number of needed certifications per year) (across the utility).
- *Employee evaluation results* (assumes utility evaluates employee performance in a routine way and documents results): Results of employee evaluations (e.g., employee growth not clearly demonstrated, employee growth only demonstrated in certain areas or for certain labor categories, etc.).
- *Presence of employee-focused objectives and targets*: Do you have employee-focused organizational objectives and targets and a related professional management system in place? Are you meeting your targets (yes/no)? (Targets could be, for instance, related to quantity, quality, timeliness, or cost. A timeliness target could, for example, relate to the number of hours it takes on average to complete a routine task.)

3. Workforce succession preparedness

Description: This measure assesses utility long-term workforce succession planning efforts to ensure critical skills and knowledge are retained and enhanced over time, particularly in light of anticipated retirement volume in coming years. Focus is on preparing entire groups or cohorts for needed workforce succession, including continued training and leadership development.

Example calculations:

- *Key position vacancies*: Average time that critical-skill positions are vacant due to staff departures per vacancy per year.
- *Key position internal/external recruitment* (percent): 100 X (number of critical-skill positions that are filled internally (through promotion, transfer, etc. rather than outside recruitment) versus filled through outside recruitment ÷ total number of positions filled per year). (This will help the utility to understand if internal workforce development is covering long-term succession needs.)

---

[7] From AWWA and AwwaRF, *Selection and Definition of Performance Indicators for Water and Wastewater Utilities*, p. 38. 2004. Note: This material is copyrighted and any reprinting must be by permission of the American Water Works Association.

- *Long-term succession plan coverage* (percent): 100 X (number of employees (or cohorts, work units, etc.) covered by a long-term workforce succession plan that accounts for projected retirements and other vacancies in each skill and management area ÷ total number of employees) (or cohorts, work units, etc.).

Operational Optimization

1. Resource optimization

Description: This measure examines resource use efficiency, including labor and material per unit of output or mile of collection/distribution system.

Example calculations:
- *Customer accounts per employee:* Number of accounts ÷ number of FTEs. (FTE = 2,080 hours per year of employee time equivalent.) This is a QualServe Indicator.[8]
- *MGD water delivered/processed per employee:* Average MGD delivered/processed ÷ FTEs per year. This is a QualServe Indicator.[9]
- *Chemical use per volume delivered/processed:* Amount of chemicals used ÷ MG delivered/processed during reporting period. (Alternatively can use dollar amount spent on chemicals ÷ MG delivered/processed; in this case a rolling average for amount spent would account for periodic bulk purchases.)
- *Energy use per volume delivered/processed:* KWH ÷ MG delivered/processed during reporting period. (Alternatively can use dollar amount spent on energy ÷ MG delivered/processed.)
- *O&M cost per volume delivered/processed:* Total O&M cost ÷ MG delivered/processed during reporting period.

A utility can also apply the above resource use per volume delivered/processed calculations to resource use per mile (or 100 miles) of collection/distribution system, (i.e., chemical use per mile, energy use per mile, or O&M cost per mile).

2. Water management efficiency

Description: This measure assesses drinking water production and delivery efficiency by considering resources as they enter and exit the utility system.

---

[8] Part of the same Indicator (set) as MGD water delivered/MGD waste water processed per FTE. From AWWA and AwwaRF, *Selection and Definition of Performance Indicators for Water and Wastewater Utilities*, p. 40. 2004. Note: This material is copyrighted and any reprinting must be by permission of the American Water Works Association.
[9] Ibid., p. 40.

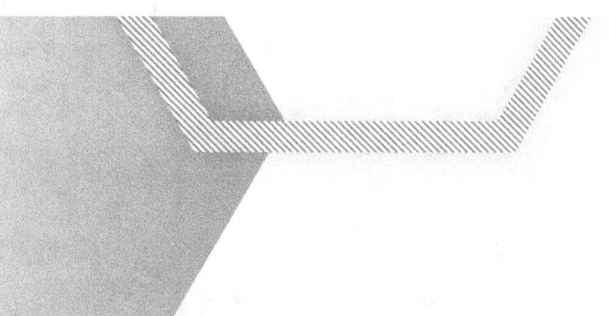

Example calculations:

- *Production efficiency*: Ratio of raw water volume taken into the treatment system to treated water produced.
- *Distribution system water loss* (a.k.a. non-revenue water) (percent): 100 X [volume of water distributed – (volume of water billed + volume of unbilled authorized water) ÷ total volume of water distributed]. (Quantifies the percentage of produced water that fails to reach customers and cannot otherwise be accounted for through authorized usage.) This is a QualServe Indicator.[10]
- *Meter function* (percent): 100 X (total number of active billable meters minus stopped or malfunctioning meters ÷ total number of active billable meters).

Financial Viability

1. Budget management effectiveness

Description: This measure has short-term and long-term aspects. The short-term calculations are commonly used financial performance indicators, and the long-term consideration is a more comprehensive analytical approach to assessing budget health over the course of several decades.

Example calculations:

Short-term (typically per year):
- *Revenue to expenditure ratio*: Total revenue ÷ total expenditures.
- *O&M expenditures* (percent): 100 X (O&M expenditures ÷ total operating budget).
- *Capital expenditures* (percent): 100 X (capital expenditures ÷ total capital budget).
- *Debt ratio*: Total liabilities ÷ total assets. Total liabilities are the entire obligations of the utility under law or equity. Total assets are the entire resource of the utility, both tangible and intangible. Utilities often have different debt-risk acceptability levels, thus the ratio itself should be considered within each utility's unique circumstances. This is a QualServe Indicator.[11]

---

[10] From AWWA and AwwaRF, *Selection and Definition of Performance Indicators for Water and Wastewater Utilities*, p. 59. 2004. Note: This material is copyrighted and any reprinting must be by permission of the American Water Works Association.
[11] Ibid., p. 51. 2004.

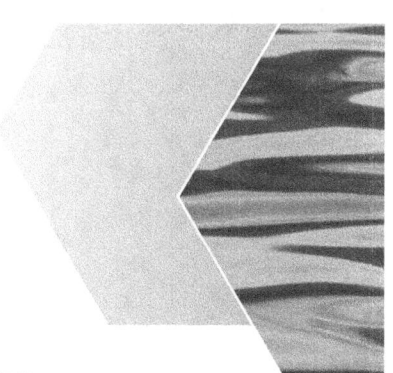

Long-term:
- *Life-cycle cost accounting:* Has the utility conducted a life-cycle cost accounting analysis[12] that explicitly incorporates accepted service level risks, asset condition, budget needs based on the values (net present values) of utility current and future assets, etc., and made financial and budget management decisions accordingly (yes/no)?

2. Financial procedure integrity

Description: Questions that gauge presence of internal utility processes to ensure a high level of financial management integrity.

Example calculations:
- Does the utility have financial accounting policies and procedures (yes/no)?
- Are financial results and internal controls audited annually (yes/no)?
- Have the number of control deficiencies and material weaknesses been reduced from previous audits (yes/no)?

3. Bond ratings

Description: Bond ratings are a general indicator of financial viability; however, they are not always within a utility's control and are less important if a utility is not participating in capital markets. Smaller utilities often struggle to obtain high ratings. Even though a higher bond rating is desirable and this provides a general indicator of financial health, the bond rating should not be considered alone. It should be considered in light of other factors such as the other measures suggested for this Attribute.

Example question:
- Has your bond rating changed recently? If so, why? Does the change reflect the utility's financial management in a way that can and should be acknowledged and, if need be, addressed?

---

[12] Section 707 of Executive Order 13123 defines life-cycle costs as, "...the sum of present values of investment costs, capital costs, installation costs, energy costs, operating costs, maintenance costs, and disposal costs over the life-time of the project, product, or measure." Life-cycle cost analysis (LCCA) is an economic method of project evaluation in which all costs arising from owning, operating, maintaining, and disposing of a [facility/asset] are considered important to the decision. LCCA is particularly suited to the evaluation of design alternatives that satisfy a required performance level, but that may have differing investment, operating, maintenance, or repair costs; and possibly different life spans. LCCA can be applied to any capital investment decision, and is particularly relevant when high initial costs are traded for reduced future cost obligations. See also: http://www.epa.gov/EMS/position/eo13148.htm, http://www.wbdg.org/resources/lcca.php.

4. Rate adequacy

Description: This measure helps the utility to consider its rates relative to factors such as external economic trends, short-term financial management, and long-term financial health. It recognizes that a "one size fits all" calculation would not be realistic due to each utility's unique situation and the number of variables that could reasonably be considered. The following three questions prompt assessment of key components of rate adequacy.

Example questions:

- How do your rate changes compare currently and over time with the inflation rate and the Consumer Price Index (CPI) or Consumer Price Index for All Urban Consumers (CPI-U)? (Rate increases below CPI for very long may suggest rates are not keeping up with utility costs.) (Using a rolling rate average over time will adjust for short-term rate hikes due to capital or O&M spending needs.)
- Have you established rates that fully consider the full life-cycle cost of service and capital funding options? (See the life-cycle cost accounting discussion, above.)
- Does your utility maintain a rate stabilization reserve to sustain operations during cycles of revenue fluctuation, in addition to 60- (or 90-) day operating reserves?

Infrastructure Stability

1. Asset inventory

Description: This measure gauges a utility's efforts to assess assets and asset conditions, as the first steps towards building a comprehensive asset management program.

Example calculations:

- *Inventory coverage* (percent): 100 X (total number of critical assets inventoried within a reasonable period of time (e.g., 5-10 years) ÷ total number of critical assets). A utility will need to first define what it considers to be a critical asset and a complete inventory will involve understanding the following for each:
  • Age and location;
  • Asset size and/or capacity;
  • Valuation data (e.g., original and replacement cost);
  • Installation date and expected service life;

- Maintenance and performance history; and
- Construction materials and recommended maintenance practices.[13]

○ *Condition assessment coverage* (percent): 100 X (total number of critical assets with condition assessed and categorized into condition categories within a reasonable period of time (e.g., 5-10 years) ÷ total number of critical assets). Condition categories could include: unacceptable, improvement needed, adequate, good, and excellent to reflect expected service levels and accepted risks.

2. Asset (system) renewal/replacement

Description: This measure assesses asset renewal/replacement rates over time. The measure should reflect utility targets, which will vary depending on each utility's determinations of acceptable risks for different asset classes. Decisions on asset replacement typically factor in internally agreed-upon risks and objectives, which may differ by asset class and other considerations. For instance, a utility may decide to run certain assets to failure based on benefit-cost analysis.

Example calculations:

○ *Asset renewal/replacement rate* (percent): 100 X (total number of assets replaced per year for each asset class ÷ total number of assets in each asset class). For example, a two percent per year replacement target (50-year renewal) for a particular asset class could be identified as the basis for performance monitoring.

— or —

○ *Asset (system) renewal/replacement rate*: 100 X (total actual expenditures or total amount of funds reserved for renewal and replacement for each asset group ÷ total present worth for renewal and replacement needs for each asset group). This is a QualServe Indicator.[14]

3. Water distribution/collection system integrity

Description: For drinking water utilities, this measure quantifies the number of pipeline leaks and breaks. Distribution system integrity has importance for health, customer service, operational, and asset management reasons. For wastewater utilities, this measure examines the frequency of collection system failures. When tracked over time, a utility can evaluate whether its failure rate is decreasing, stable, or increasing. When data are maintained to characterize failures by pipe type and age, type

---

[13] From the U.S. General Accounting Office, *Water Infrastructure: Comprehensive Asset Management Has Potential to Help Utilities Better Identify Needs and Plan Future Investments.* GAO-04-461. March 2004. Available: http://www.gao.gov/new.items/d04461.pdf.

[14] From AWWA and AwwaRF, *Selection and Definition of Performance Indicators for Water and Wastewater Utilities,* p. 53. 2004. Note: This material is copyrighted and any reprinting must be by permission of the American Water Works Association.

of failure, and cost of repairs, decisions regarding routine maintenance and replacement/renewals can be better made.[15]

Example calculation (drinking water utilities):

- *Leakage and breakage frequency rate* (percent): 100 X ((total number of leaks + total number of breaks) ÷ total miles of distribution piping per year). (Note: leaks and breaks are distinctly different events.) This is a QualServe Indicator.[16]

Example calculation (wastewater utilities):

- *Collection system failure rate* (percent): 100 X (total number of collection system failures ÷ total miles of collection system piping per year). This is a QualServe Indicator.[17]

4. Planned maintenance

Description: Planned maintenance includes both preventive and predictive maintenance. Preventive maintenance is performed according to a predetermined schedule rather than in response to failure. Predictive maintenance is initiated when signals indicate that maintenance is due. All other maintenance is categorized as corrective or reactive.[18]

Example calculations:

This measure can be measured in different ways. Calculating costs may be preferable to encourage business decisions based on total cost; however, the reliability of costs is uncertain. Hours are likely to be less variable than costs, but not all utilities track hours. Thus, cost and hours ratios are desirable, where possible.

- *Planned maintenance ratio by hours* (percent): 100 X (hours of planned maintenance ÷ (hours of planned + corrective maintenance)). This is a QualServe Indicator.[19]
- *Planned maintenance ratio by cost* (percent): 100 X (cost of planned maintenance ÷ (cost of planned + corrective maintenance)). This is a QualServe Indicator.[20]

---

[15] From AWWA and AwwaRF, *Selection and Definition of Performance Indicators for Water and Wastewater Utilities*, p. 70. 2004. Note: This material is copyrighted and any reprinting must be by permission of the American Water Works Association.
[16] Ibid., p. 61.
[17] Ibid., p. 70.
[18] Ibid., p. 65.
[19] Ibid., p. 66.
[20] Ibid., p. 66.

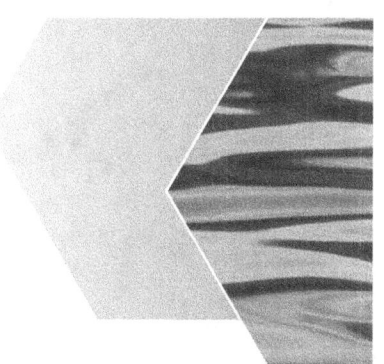

Operational Resiliency

1. Recordable incidents of injury or illnesses

Description: Incidence rates can be used to show the relative level of injuries and illnesses and help determine problem areas and progress in preventing work-related injuries and illnesses.

Example calculations:

The U.S. Bureau of Labor Statistics has developed instructions for employers to evaluate their firm's injury and illness record. The calculation below is based on these instructions, which can be accessed at: http://www.bls.gov/iif/osheval.htm.

- *Total recordable incident rate*: (Number of work-related injuries and illnesses X 200,000[21]) ÷ employee hours worked.

2. Insurance claims

Description: This measure examines the number, type, and severity of insurance claims to understand insurance coverage strength/vulnerability.

Example calculations:

- *Number of insurance claims*: Number of general liability and auto insurance claims per 200,000[22] employee hours worked.
- *Severity of insurance claims*: Total dollar amount of general liability and auto insurance claims per 200,000[23] employee hours worked.

3. Risk assessment and response preparedness

Description: This measure asks whether utilities have assessed their all-hazards (natural and human-caused) vulnerabilities and risks and made corresponding plans for critical needs. Risk assessment in this context includes a vulnerability assessment regarding, for example, power outages, lack of access to chemicals, curtailed staff availability, etc.

---

[21] 200,000 hours is a standard number used by OSHA to normalize data. It represents the equivalent of 100 employees working 40 hours per week, 50 weeks per year, and provides the standard base for the incidence rates.
[22] See the explanation in the footnote above regarding the 200,000 hours standard.
[23] See the explanation in the footnote above regarding the 200,000 hours standard.

Example calculations:

- Emergency Response Plan (ERP) coverage and preparedness:
  - Does the utility have an ERP in place (yes/no)?
  - Number and frequency of ERP trainings per year: 100 X (number of employees who participate in ERP trainings ÷ total number of employees).
  - Number and frequency of ERP exercises per year: 100 X (number of employees who participate in ERP exercises ÷ total number of employees).
  - Frequency with which the ERP is reviewed and updated.
- *Vulnerability management*: Is there a process in place for identifying and addressing system deficiencies (e.g., deficiency reporting with an immediate remedy process) (yes/no)?

4. Ongoing operational resiliency

Description: This measure assesses a utility's operational reliability during ongoing/routine operations.

Example calculations:

- *Uptime for critical utility components on an ongoing basis* (percent): 100 X (hours of critical component uptime ÷ hours critical components have the physical potential to be operational). Note: a utility can apply this measure on an individual component basis or summed across all identified critical components. Also, a utility can make this measure more precise by adjusting for planned maintenance periods.

5. Operational resiliency under emergency conditions

Description: This measure assesses the operational preparedness and expected responsiveness in critical areas under emergency conditions.

Example calculations (all apply to emergency conditions and, where relevant, factor in anticipated downtimes relative to required/high demand times):

- *Power resiliency*: Period of time (e.g., hours or days) for which backup power is available for critical operations (i.e., those required to meet 100 percent of minimum daily demand). (Note: "minimum daily demand" is the average daily demand for the lowest production month of the year.)
- *Treatment chemical resiliency*: Period of time (e.g., hours or days) minimum daily demand can be met with water treated to meet SDWA standards for acute contaminants (i.e., *E.coli*, fecal coliform, nitrate, nitrite, total nitrate and nitrite, chlorine dioxide, turbidity as referenced in the list of situations requiring a Tier 1 Public Notification under 40 CFR 141.202), without additional treatment

chemical deliveries. (Note: "minimum daily demand" is the average daily demand for the lowest production month of the year.)

- *Critical parts and equipment resiliency*: Current longest lead time (e.g., hours or days) for repair or replacement of operationally critical parts or equipment (calculated by examining repair and replacement lead times for all identified critical parts and equipment and taking the longest single identified time).
- *Critical staff resiliency*: Average number of response-capable backup staff for critical operation and maintenance positions (calculated as the sum of all response-capable backup staff ÷ total number of critical operation and maintenance positions).
- *Treatment operations resiliency* (percent): Percent of minimum daily demand met with the primary production or treatment plant offline for 24, 48, and 72 hours. (Note: "minimum daily demand" is the average daily demand for the lowest production month of the year.)
- *Sourcewater resiliency*: Period of time (e.g., hours or days) minimum daily demand can be met with the primary raw water source unavailable. (Note: "minimum daily demand" is the average daily demand for the lowest production month of the year.)

Community Sustainability

1. Watershed-based infrastructure planning

Description: This measure addresses utility efforts to consider watershed-based approaches when making management decisions affecting infrastructure planning and investment options. Watershed protection strategies can sometimes, for example, protect sourcewater quality limiting the need for additional or enhanced water treatment capacity.

Example question:

- Does the utility employ alternative, watershed-based approaches to align infrastructure decisions with overall watershed goals and potentially reduce future infrastructure costs? Watershed-based approaches include, for example: centralized management of decentralized systems; stormwater management; sourcewater protection programs; and conjunctive use of groundwater, sourcewater, and recycled water to optimize resource use at a basin scale. (See also "green infrastructure" below.)

2. Green infrastructure

Description: "Green infrastructure" includes both the built and natural/unbuilt environment. Utilities may promote source water protection and conservation "green infrastructure" approaches in support of water conservation (e.g., per capita demand reduction) and water quality protection objectives. Green infrastructure approaches can include: low-impact development techniques (e.g., minimization of impervious surfaces, green roofs); protection of green spaces and wildlife habitat; incentives for water-efficient domestic appliance use and landscaping; green building standards such as those promoted through the Leadership in Energy and Environmental Design (LEED) program; management of energy, chemical, and material use; etc.[24] Utilities often coordinate these efforts with community planning offices.

Example question:

- Has the utility explored green infrastructure approaches and opportunities that are aligned with the utility's mandate, goals, and objectives and community interests (yes/no)?
- Does the utility have procedures that incorporate green infrastructure approaches and performance into new infrastructure investments (yes/no)?

3. Greenhouse gas emissions

Description: This measure will help drinking and wastewater utilities to understand and reduce their individual contributions to area greenhouse gas emissions. Trends indicate that water utility emissions of these gases will likely be of interest to stakeholders. Monitoring of these emissions is becoming more common among water sector utilities, and some utilities are beginning voluntary efforts to reduce their emissions (e.g., through production of reusable methane energy by wastewater utilities).

Example calculation:

- Net (gross minus offsets) greenhouse gas emissions in tons of carbon dioxide (CO_2), nitrous oxide (N_2O), methane (CH_4), and, as applicable, hydrofluorocarbons (HFCs) and perfluorocarbons (PFCs). Start by establishing an emissions baseline and then track emission trends in conjunction with minimizing/reducing emissions over time, where possible.[25] Emissions inventories often incorporate indirect emissions such as those generated during the production and transport of materials and chemicals.

---

[24] For more information about green infrastructure, visit www.epa.gov/npdes/greeninfrastructure.
[25] EPA's industry-government "Climate Leaders" partnership involves completing a corporate-wide inventory of their greenhouse gas emissions. Information and related guidance is available at http://www.epa.gov/stateply/index.html.

4. Service affordability

Description: Drinking water and wastewater service affordability centers on community members' ability to pay for water services. The true cost of water/wastewater services may be higher than some low-income households can afford, particularly when rates reflect the full life-cycle cost of water services. Each utility will want to consider and balance keeping water services affordable while ensuring the rates needed for long-term infrastructure and financial integrity.

Example calculations and considerations:

- *Bill affordability* (households for which rates may represent an unaffordable level) (percent): 100 X (number of households served for which average water bill is > "X" percent (often 2-2.5%) of median household income[26] ÷ total number of households served).

Coupled with:

- *Low-income billing assistance program coverage* (percent): 100 X (number of customers enrolled in low-income billing assistance program ÷ number of customers who are eligible for enrollment in low-income billing assistance program). (The utility can try to increase participation in the program for eligible households that are not participating.)

Water Resource Adequacy

1. Water supply adequacy

Description: This measure assesses short-term and long-term water supply adequacy and explores related long-term supply considerations.

Example calculations and questions:

- *Short-term water supply adequacy*: Period of time for which existing supply sources are adequate. This can be measured as a ratio of projected short-term (e.g., 12-month rolling average) monthly supply to projected short-term monthly demand. Often an index or scale is used, for example, short-term supply relative to severe drought (assigned a "1") to abundant supply conditions (assigned a "5").

---

[26] This calculation focuses on identifying low-income households based median household incomes (MHI); however, MHI is not strongly correlated with the incidence of poverty or other measures of economic need. Further, populations served by small utilities in rural settings tend to have lower MHI and higher poverty rates, but fewer options for diversifying water/wastewater service rates based on need compared to larger municipal systems.

○ *Long-term water supply adequacy*: Projected future annual supply relative to projected future annual demand for at least the next 50 years (some utilities project out as far as 70-80 years). Statistical forecasting and simulation modeling and forecasting techniques are typically used for such long-term projections. Analysis variables in addition to historical record (e.g., historical and year-to-date reservoir elevation data), forecasted precipitation, and flows can include:
  - Future normal, wet, dry, and very dry scenarios (including anticipated climate change-related scenarios);
  - Anticipated population changes;
  - Future service areas;
  - Availability of new water supplies, including recycled water (plus availability of water rights for new supplies, where applicable); and
  - Levels of uncertainty around the above.

2. Supply and demand management

Description: This metric explores whether the utility has a strategy for proactive supply and demand management in the short and long terms. Strategy needs will depend on community circumstances and priorities, anticipated population growth, future water supply in relation to anticipated demand, demand management and other conservation options, and other local considerations.

Example questions:

○ Has the utility developed a sourcewater protection plan (yes/no) and is the plan current (yes/no)?

○ Does the utility have a demand management/demand reduction plan (yes/no)? Does this plan track per capita water consumption and, where analytical tools are available to do so, accurately attribute per capita consumption reductions to demand reduction strategies (such as public education and rebates for water-efficient appliances) (yes/no)?

○ Do demand scenarios account for changes in rates (which can change for many reasons) and conservation-oriented, demand management pricing structures (yes/no)?

○ Does the utility have policies in place that address, prior to committing to new service areas, availability of adequate dry year supply (yes/no)? Alternatively, does the utility have a commitment to denying service commitments unless a reliable drought-year supply, with reasonable drought use restrictions, is available to meet the commitment (yes/no)?

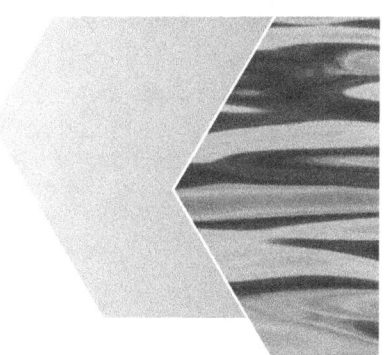

Stakeholder Understanding and Support

1. Stakeholder consultation

Description: This measure addresses utility actions to reach out to and consult with stakeholders about utility matters, including utility goals, objectives, and management decisions.

Example questions:

- Does the utility identify stakeholders, conduct outreach, and actively consult with stakeholders about utility matters (yes/no)? Elements of this plan can include:
  - Number of active contacts with stakeholders in key areas (e.g., from local government, business, education, non-governmental groups)?
  - Does the utility actively seek input from stakeholders (yes/no)?
  - Frequency with which the utility actively consults with stakeholders. This measure should go beyond counting the number of calls or times information is sent out or posted on websites to items such as number of stakeholder outreach and education activities, number of opportunities for stakeholders to provide input, participation of stakeholders on utility committees, etc.
- Does the utility actively consider and act upon stakeholder input (yes/no)?

2. Stakeholder satisfaction

Description: This measure addresses stakeholder perceptions of the utility. Stakeholder satisfaction can be measured through surveys sent to stakeholders, formal feedback surveys distributed to stakeholders at events, etc.

Example calculations:

- *Overall satisfaction* (percent): 100 X (number of stakeholders who annually rate the overall job of the utility as positive ÷ total number of stakeholders surveyed).
- *Responsiveness* (percent): 100 X (number of stakeholders who annually rate utility responsiveness to stakeholder needs as positive ÷ total number of stakeholders surveyed).
- *Message recollection for outreach programs targeted to specific stakeholder groups* (percent): (a) 100 X (number of stakeholders who recall key messages ÷ total number of stakeholders surveyed); and (b) 100 X (number of stakeholders who recall the message source (TV, utility mailers, newsletters, etc.) ÷ total number of stakeholders surveyed).

3. Internal benefits from stakeholder input

Description: This measure addresses the value utility employees believe stakeholder engagement has provided to utility projects and activities. Measurement by the utility can focus on surveying utility employees running projects that have stakeholder involvement.

Example calculations:

- 100 X (number of utility projects or activities where stakeholders participated and/or provided input for which utility employees believe there was value added as a result of stakeholder participation and input ÷ total number of projects where stakeholders participated and/or provided input).
- *Overall value added* (percent): 100 X (number of utility employees who rated their overall sense of value added from stakeholder participation and input as (high value added, some value added, little value added, no value added) ÷ total number of utility employees surveyed).

4. Comparative rate rank

Description: This measure depicts how utility rates compare to similar utilities (e.g., utilities of the same type (drinking water, wastewater) that are similar in terms of geographic region, size of population served, etc.). A utility can use the measure internally or to educate stakeholders. It should be noted that the lowest rate is not necessarily best (see Financial Viability).

Example calculations:

- Typical monthly bill for the average household as a percentage of typical monthly bills for similar area utilities.

5. Media/press coverage

Description: This measure captures media portrayal of the utility (newspaper, TV, radio, etc.) in terms of awareness, accuracy, and tone.

Example calculations:

- *Amount of coverage*: Total number of media stories (newspaper, TV, radio, etc.) concerning the utility per year.
- *Media coverage tone* (percent): 100 X (number of media stories concerning the utility that portray the utility in a positive way ÷ total number of media stories concerning the utility) per year.
- *Media coverage accuracy* (percent): 100 X (number of media stories that accurately describe the utility ÷ total number of media stories concerning the utility) per year.

www.ingramcontent.com/pod-product-compliance
Lightning Source LLC
Chambersburg PA
CBHW081751170526
45167CB00009B/4001